© Houghton Mifflin Harcourt Publishing Company • Cover Image Credits: (Hares) ©Radius Images/Corbis; (Garden, New York) ©Rick Lew/The Image Bank/Getty Images; (sky) ©PhotoDisc/Getty Images

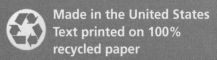

Made in the United States
Text printed on 100%
recycled paper

**Houghton
Mifflin
Harcourt**

Copyright © 2015 by Houghton Mifflin Harcourt Publishing Company

All rights reserved. No part of this work may be reproduced or
transmitted in any form or by any means, electronic or mechanical,
including photocopying or recording, or by any information
storage and retrieval system, without the prior written permission of
the copyright owner unless such copying is expressly permitted by
federal copyright law. Requests for permission to make copies of any
part of the work should be addressed to Houghton Mifflin Harcourt
Publishing Company, Attn: Contracts, Copyrights, and Licensing, 9400
Southpark Center Loop, Orlando, Florida 32819-8647.

Printed in the U.S.A.

ISBN 978-0-544-34172-2

3 4 5 6 7 8 9 10 0877 22 21 20 19 18 17 16 15 14

4500499660   ^ B C D E F G

If you have received these materials as examination copies free of
charge, Houghton Mifflin Harcourt Publishing Company retains title
to the materials and they may not be resold. Resale of examination
copies is strictly prohibited.

Possession of this publication in print format does not entitle users
to convert this publication, or any portion of it, into electronic
format.

Dear Students and Families,

Welcome to **Go Math!**, Grade K! In this exciting mathematics program, there are hands-on activities to do and real-world problems to solve. Best of all, you will write your ideas and answers right in your book. In **Go Math!**, writing and drawing on the pages helps you think deeply about what you are learning, and you will really understand math!

By the way, all of the pages in your **Go Math!** book are made using recycled paper. We wanted you to know that you can Go Green with **Go Math!**

Sincerely,

The Authors

Made in the United States
Text printed on 100% recycled paper

© Houghton Mifflin Harcourt Publishing Company • Image Credits: (bg) ©Sankar Salvady/Flickr/Getty Images; (t) ©Blaine Harrington III/Alamy Images; (c) ©Don Johnston/All Canada Photos/Getty Images; (b) ©Erich Kuchling/Westend61/Corbis

Authors

Juli K. Dixon, Ph.D.
Professor, Mathematics Education
University of Central Florida
Orlando, Florida

Edward B. Burger, Ph.D.
President, Southwestern University
Georgetown, Texas

Steven J. Leinwand
Principal Research Analyst
American Institutes for
   Research (AIR)
Washington, D.C.

Contributor

Rena Petrello
Professor, Mathematics
Moorpark College
Moorpark, California

Matthew R. Larson, Ph.D.
K-12 Curriculum Specialist for
   Mathematics
Lincoln Public Schools
Lincoln, Nebraska

Martha E. Sandoval-Martinez
Math Instructor
El Camino College
Torrance, California

English Language Learners Consultant

Elizabeth Jiménez
CEO, GEMAS Consulting
Professional Expert on English
   Learner Education
Bilingual Education and
   Dual Language
Pomona, California

© Houghton Mifflin Harcourt Publishing Company • Image Credits: (bg) ©Russ Bishop/Alamy Images ; (t) ©Richard Wear/Design Pics/Corbis

Number and Operations

Common Core **Critical Area** Representing, relating, and operating on whole numbers, initially with sets of objects.

 Represent and Compare Numbers to 10   **177**

COMMON CORE STATE STANDARDS

K.CC Counting and Cardinality
Cluster A Know number names and count the sequence.
K.CC.A.2, K.CC.A.3
Cluster B Count to tell the number of objects.
K.CC.B.5
Cluster C Compare numbers.
K.CC.C.6, K.CC.C.7
K.OA Operations and Algebraic Thinking
Cluster A Understand addition as putting together and adding to, and understand subtraction as taking apart and taking from.
K.OA.A.4

GO DIGITAL

Go online! Your math lessons are interactive. Use *i*Tools, Animated Math Models, the Multimedia *e*Glossary, and more.

Chapter 4 Overview

In this chapter, you will explore and discover answers to the following **Essential Questions**:

• How can you show and compare numbers to 10?
• How can you count forward to 10?
• How can you show numbers from 1 to 10?
• How can using models help you compare two numbers?

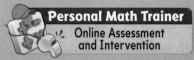
Personal Math Trainer
Online Assessment and Intervention

© Houghton Mifflin Harcourt Publishing Company

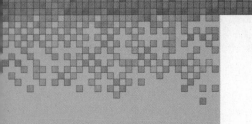

FOR MORE PRACTICE
GO TO THE
Personal Math Trainer

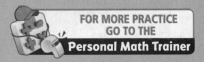

Practice and Homework

Lesson Check and
Spiral Review in
every lesson

© Houghton Mifflin Harcourt Publishing Company

Represent and Compare Numbers to 10

Curious About Math with Curious George

Apple trees grow from a small seed.

• About how many seeds are in an apple?

© Houghton Mifflin Harcourt Publishing Company • Image Credits: (bg) ©Herbert Kehrer/Corbis
Curious George by Margret and H.A. Rey. Copyright © 2010 by Houghton Mifflin Harcourt Publishing Company.
All rights reserved. The character Curious George®, including without limitation the character's name and the
character's likenesses, are registered trademarks of Houghton Mifflin Harcourt Publishing Company.

Name _____

 **Show What You Know**

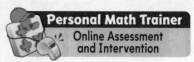
Personal Math Trainer
Online Assessment
and Intervention

Draw Objects to 9

 1

9

 2

7

Write Numbers to 9

3  _____ - - - - - _____

4  _____ - - - - - _____

5  _____ - - - - - _____

6 _____ - - - - - _____

This page checks understanding of important skills needed for success in Chapter 4.

DIRECTIONS 1. Draw 9 flowers. **2.** Draw 7 flowers.
3–6. Count and tell how many. Write the number.

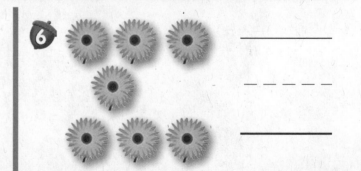

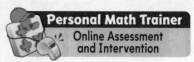

© Houghton Mifflin Harcourt Publishing Company

Name _____

Vocabulary Builder

greater

less

same number

DIRECTIONS Circle the words that describe the number of carrots and the number of celery sticks. Use *greater* and *less* to describe the number of trees and the number of bushes.

GO DIGITAL
• **Interactive Student Edition**
• **Multimedia eGlossary**

© Houghton Mifflin Harcourt Publishing Company

Chapter 4

Game

Spin and Count!

START

END

DIRECTIONS Play with a partner. Place game markers on START. Use a pencil and a paper clip to spin for a number. Take turns spinning. Each player moves his or her marker to the next space that has the same number of objects as the number on the spinner. The first player to reach END wins.

MATERIALS two game markers, pencil, paper clip

© Houghton Mifflin Harcourt Publishing Company

Chapter 4 Vocabulary

and

y

4

compare

comparar

13

greater

mayor

31

less

menor, menos

38

match

emparejar

41

pairs

pares

50

same number

el mismo número

57

ten

diez

74

The number of blue counters **compares** equally to the number of red counters.

© Houghton Mifflin Harcourt Publishing Company

 **and**

2 + 2

© Houghton Mifflin Harcourt Publishing Company

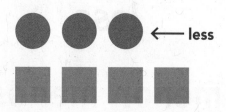 ← **less**

© Houghton Mifflin Harcourt Publishing Company

 6

9

9 red cubes is **greater**

© Houghton Mifflin Harcourt Publishing Company

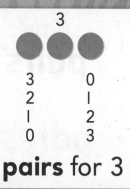

3

3   0
2   1
1   2
0   3

pairs for 3

© Houghton Mifflin Harcourt Publishing Company

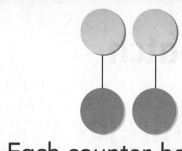

Each counter has a **match**.

© Houghton Mifflin Harcourt Publishing Company

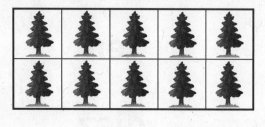

10

© Houghton Mifflin Harcourt Publishing Company

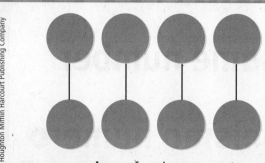

same number of red counters in each row

Memory

Word Box

and

compare

greater

less

match

pair

same number

ten

DIRECTIONS Shuffle the Word Cards. Place each card facedown on a different square above. A player turns over two cards. If they match, the player tells what they know about the word and keeps the cards. If they do not match, the player turns the cards facedown again. Players take turns. The player with more pairs wins.

MATERIALS I set of Word Cards

© Houghton Mifflin Harcourt Publishing Company

The Write Way

DIRECTIONS Draw to show how to compare two sets of objects.
Reflect Be ready to tell about your drawing.

© Houghton Mifflin Harcourt Publishing Company

Name _____

Model and Count 10

Essential Question How can you show and count 10 objects?

Common Core **Counting and Cardinality—K.CC.B.5**
Also K.CC.B.4a, K.CC.B.4b, K.CC.B.4c
MATHEMATICAL PRACTICES
MP4, MP5

Listen and Draw Real World   Hands On

DIRECTIONS Use counters to model 9 in the top ten frame. Use counters to model 10 in the bottom ten frame. Draw the counters. Tell about the ten frames.

© Houghton Mifflin Harcourt Publishing Company • Image Credits: (bg) ©PhotoDisc/Getty Images

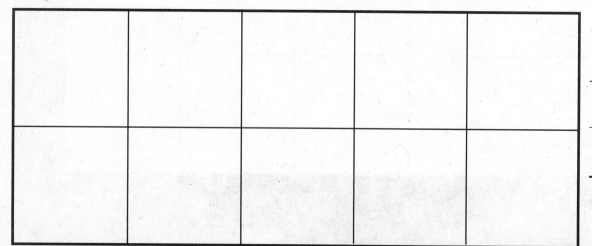

ten

DIRECTIONS 1. Place a counter on each balloon. 2. Move the counters to the ten frame. Draw the counters. Point to each counter as you count it. Trace the number.

© Houghton Mifflin Harcourt Publishing Company

Name _____

3

10
ten

_____ and _____

_____ and _____

_____ and _____

_____ and _____

© Houghton Mifflin Harcourt Publishing Company

DIRECTIONS **3.** Trace the number. Use counters to model the different ways to make 10. Write to show some pairs of numbers that make 10.

Problem Solving • Applications Real World

4

5

DIRECTIONS 4. Michelle puts her star stickers in sets of 10. Circle all the sets of star stickers that belong to Michelle. 5. Draw to show what you know about the number 10. Tell a friend about your drawing.

HOME ACTIVITY • Ask your child to show a set of nine objects. Then have him or her show one more object and tell how many objects are in the set.

© Houghton Mifflin Harcourt Publishing Company

Model and Count 10

Common Core COMMON CORE STANDARD—K.CC.B.5
Count to tell the number of objects.

ten

and

and

and

and

DIRECTIONS Trace the number. Use counters to model the different ways to make 10. Color to show the counters below. Write to show some pairs of numbers that make 10.

© Houghton Mifflin Harcourt Publishing Company

Lesson Check (K.CC.B.5)

 ten

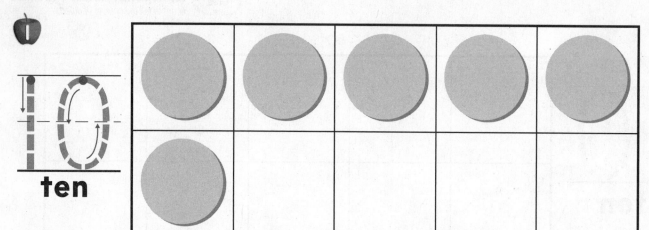

Spiral Review (K.CC.C.6, K.CC.A.3)

②

_____

— — — — — — —

_____

— — — — — — —

_____

③

_____

— — — — — — —

_____

DIRECTIONS 1. Trace the number. How many more counters would you place in the ten frame to model a way to make 10? Draw the counters. 2. Count how many kites. Write the number. Draw to show a set of counters that has the same number as the set of kites. Write the number. 3. Count and tell how many. Write the number.

© Houghton Mifflin Harcourt Publishing Company

FOR MORE PRACTICE
GO TO THE
Personal Math Trainer

Name _____

Count and Write to 10

Essential Question How can you count and write up to 10 with words and numbers?

Common Core **Counting and Cardinality—K.CC.A.3**
Also K.CC.B.4b, K.CC.B.5
MATHEMATICAL PRACTICES
MP2

Listen and Draw Real World

DIRECTIONS Count and tell how many cubes. Trace the numbers. Count and tell how many eggs. Trace the numbers and the word.

Chapter 4 • Lesson 2

© Houghton Mifflin Harcourt Publishing Company

1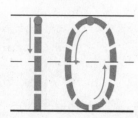

10
ten

 2

_____

- - - - - - - - - -

_____

 3

_____

- - - - - - - - - -

_____

 4 ✓

_____

- - - - - - - - - -

_____

5 ✓

_____

- - - - - - - - - -

_____

DIRECTIONS 1. Count and tell how many eggs. Trace the number. **2–5.** Count and tell how many eggs. Write the number.

© Houghton Mifflin Harcourt Publishing Company

Name _____

10
ten

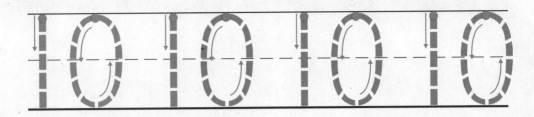

❤ 7

_____

- - - - - - - - - - - - - - -

_____

🐟 8

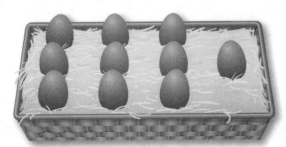

_____

- - - - - - - - - - - - - - -

_____

🐚 9

_____

- - - - - - - - - - - - - - -

_____

DIRECTIONS 6. Say the number. Trace the numbers.
7–9. Count and tell how many. Write the number.

© Houghton Mifflin Harcourt Publishing Company

Chapter 4 • Lesson 2

one hundred eighty-nine **189**

Problem Solving • Applications

WRITE
Math

10

_____

- - - - - - - - - -

_____

DIRECTIONS 10. Draw to show a set that has a number of objects one greater than 9. Write how many objects. Tell a friend about your drawing.

HOME ACTIVITY • Show ten objects. Have your child point to each object in the set as he or she counts them. Then have him or her write the number on paper to show how many objects.

190 one hundred ninety

© Houghton Mifflin Harcourt Publishing Company

Name _____

Count and Write to 10

Common Core
COMMON CORE STANDARD—K.CC.A.3
Know number names and the count sequence.

1

10
ten

10 10 10 10 10

2

_____

- - - - - - - - - -

_____

3

_____

- - - - - - - - - -

_____

4

_____

- - - - - - - - - -

_____

DIRECTIONS 1. Say the number. Trace the numbers.
2–4. Count and tell how many. Write the number.

© Houghton Mifflin Harcourt Publishing Company

Chapter 4

Lesson Check (K.CC.A.3)

 1

_____

- - - - - - - - - - - - -

_____

Spiral Review (K.CC.C.6, K.CC.B.4a)

 2

_____   _____

- - - - - - -   - - - - - - -

_____   _____

3

DIRECTIONS **1.** Count and tell how many ears of corn. Write the number. **2.** Count and tell how many are in each set. Write the numbers. Compare the numbers. Circle the number that is less. **3.** How many counters would you place in the five frame? Trace the number.

© Houghton Mifflin Harcourt Publishing Company

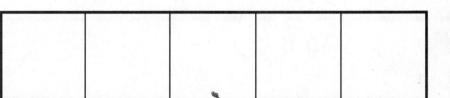

FOR MORE PRACTICE
GO TO THE
Personal Math Trainer

Name _____

Algebra • Ways to Make 10

Essential Question How can you use a drawing to make 10 from a given number?

Common Core
Operations and Algebraic Thinking—K.OA.A.4
Also K.OA.A.3
MATHEMATICAL PRACTICES
MP4, MP7

Listen and Draw

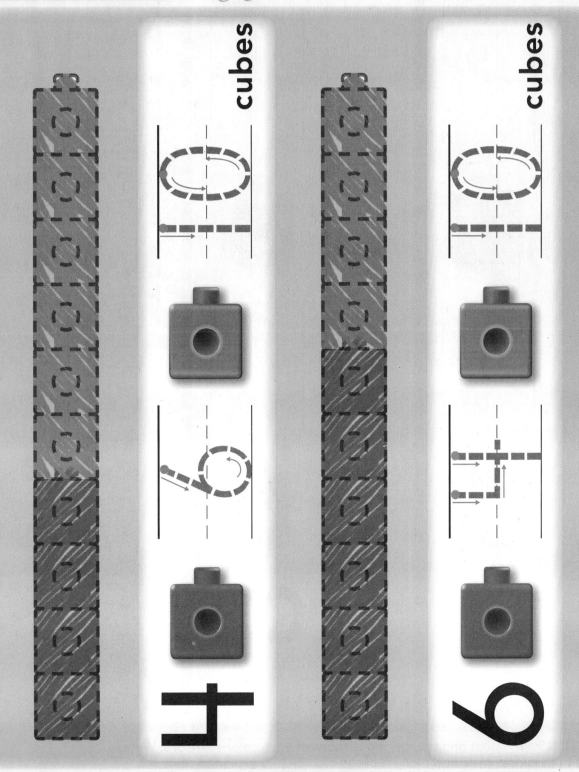

cubes

10

4

cubes

10

6

© Houghton Mifflin Harcourt Publishing Company

DIRECTIONS Use cubes of two colors to show different ways to make 10. Trace the number of red cubes. Trace the number of cubes in all.

Chapter 4 • Lesson 3

one hundred ninety-three **193**

1.

9

cubes

2.

8

cubes

3.

7

cubes

DIRECTIONS 1. Count and tell how many cubes of each color there are. Write how many red cubes. Write how many cubes in all. 2–3. Use blue to color the cubes to match the number. Use red to color the other cubes. Write how many red cubes. Write how many cubes in all.

194 one hundred ninety-four

© Houghton Mifflin Harcourt Publishing Company

Name _____

cubes

cubes

cubes

5

3

2

© Houghton Mifflin Harcourt Publishing Company

DIRECTIONS 4–6. Use blue to color the cubes to match the number. Use red to color the other cubes. Write how many red cubes. Write how many cubes in all.

Problem Solving • Applications Real World

WRITE Math

10

10

10

7

8

9

DIRECTIONS 7–9. Jill uses the dot side of two Number Tiles to make 10. Draw the dots on each Number Tile to show a way Jill can make 10. Write the numbers.

HOME ACTIVITY • Ask your child to show a set of 10 objects, using objects of the same kind that are different in one way; for example, large and small paper clips. Then have him or her write the numbers that show how many of each kind are in the set.

© Houghton Mifflin Harcourt Publishing Company

Name _____

Algebra • Ways to Make 10

Common Core

COMMON CORE STANDARD—K.OA.A.4
*Understand addition as putting together and
adding to, and understand subtraction as
taking apart and taking from.*

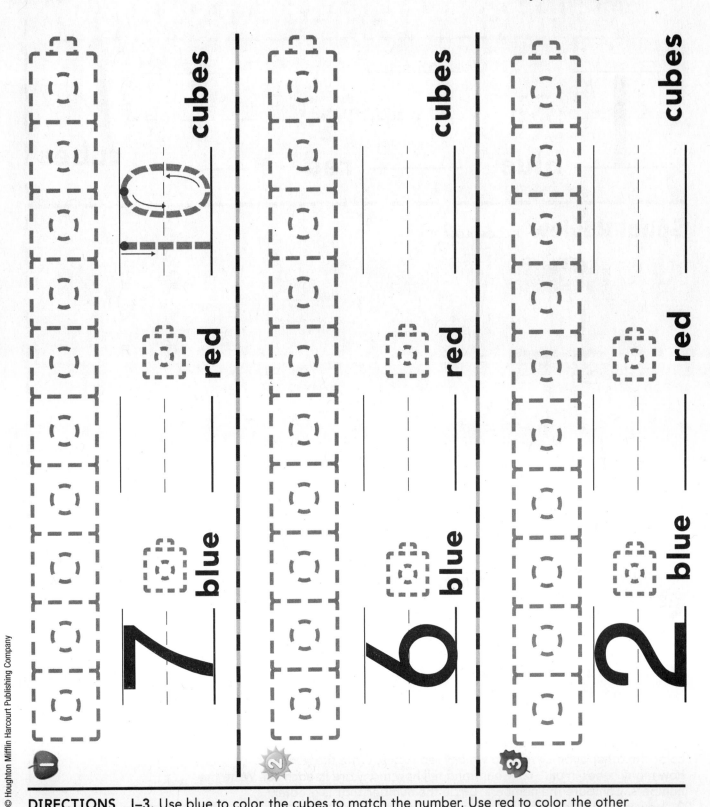

1. _____ blue _____ red 10 cubes

2. _____ blue _____ red _____ cubes

3. _____ blue _____ red _____ cubes

DIRECTIONS 1–3. Use blue to color the cubes to match the number. Use red to color the other cubes. Write how many red cubes. Trace or write the number that shows how many cubes in all.

Chapter 4

© Houghton Mifflin Harcourt Publishing Company

Lesson Check (K.OA.A.4)

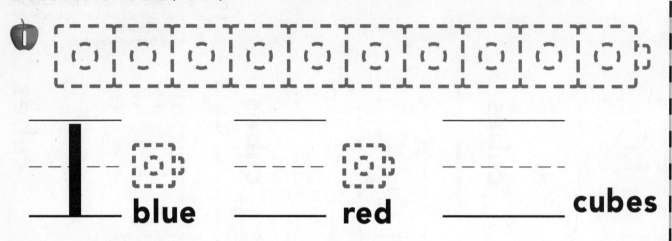

 l   _____ **blue**   _____ **red**   _____ **cubes**

Spiral Review (K.CC.C.6, K.CC.A.3)

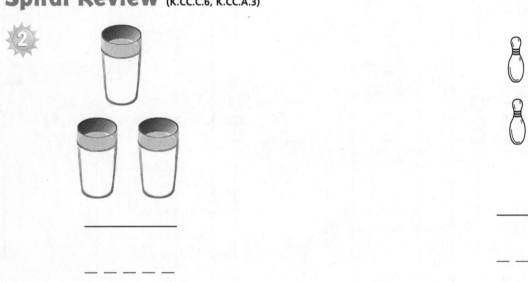

_____

- - - - -

_____

_____

- - - - -

_____

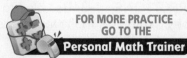

_____

- - - - -

_____

DIRECTIONS 1. Use blue to color the cube to match the number. Use red to color the other cubes. Write how many red cubes. Write the number that shows how many cubes in all.   2. Count and tell how many are in each set. Write the numbers. Compare the numbers. Circle the number that is greater.
3. How many birds are there? Write the number.

© Houghton Mifflin Harcourt Publishing Company

FOR MORE PRACTICE
GO TO THE
Personal Math Trainer

Name _____

Count and Order to 10

Essential Question How can you count forward to 10 from a given number?

Common Core **Counting and Cardinality—K.CC.A.2**
MATHEMATICAL PRACTICES
MP2

Listen and Draw

1 2 3 4 5 6 7 8 9 10

1 2 3 4 ___ 6 7 ___ 9 10

DIRECTIONS Point to the numbers in the top workspace as you count forward to 10. Trace and write the numbers in order in the bottom workspace as you count forward to 10.

Chapter 4 • Lesson 4

© Houghton Mifflin Harcourt Publishing Company • Image Credits: (bg) ©PhotoDisc/Getty Images

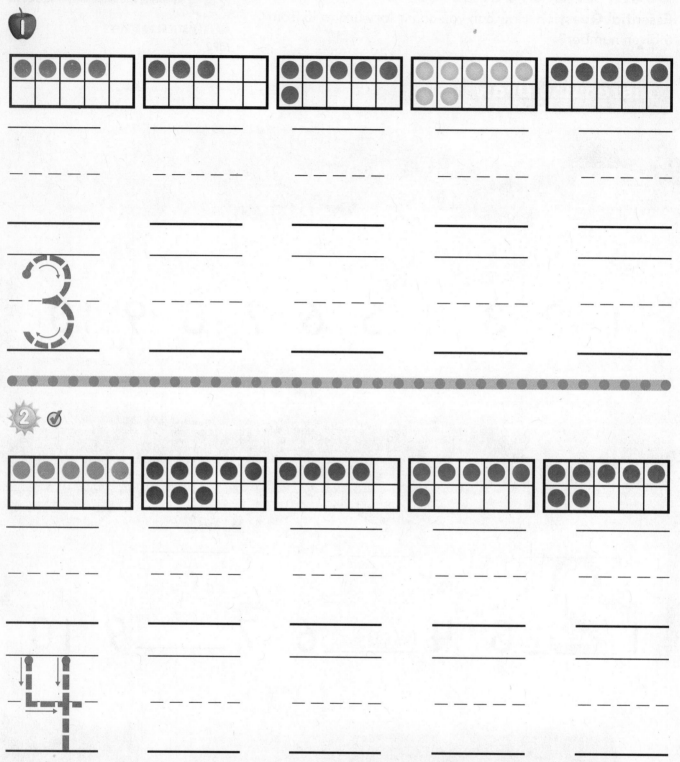

DIRECTIONS 1–2. Count the dots of each color in the ten frames. Write the numbers. Look at the next line. Write the numbers in order as you count forward from the dashed number.

© Houghton Mifflin Harcourt Publishing Company

Name _____

_____ _____ _____ _____ _____

- -

_____ _____ _____ _____ _____

5 -

_____ _____ _____ _____ _____

❹ ✅

_____ _____ _____ _____ _____

- -

_____ _____ _____ _____ _____

6 -

_____ _____ _____ _____ _____

DIRECTIONS **3–4.** Count the dots of each color in the ten frames. Write the numbers. Look at the next line. Write the numbers in order as you count forward from the dashed number.

HOME ACTIVITY • Write the numbers 1 to 10 in order on a piece of paper. Ask your child to point to each number as he or she counts to 10. Repeat beginning with a number other than 1 when counting.

© Houghton Mifflin Harcourt Publishing Company

Concepts and Skills

🍎 **1**

_____

_ _ _ _ _

_____

• •

⚙ **2**

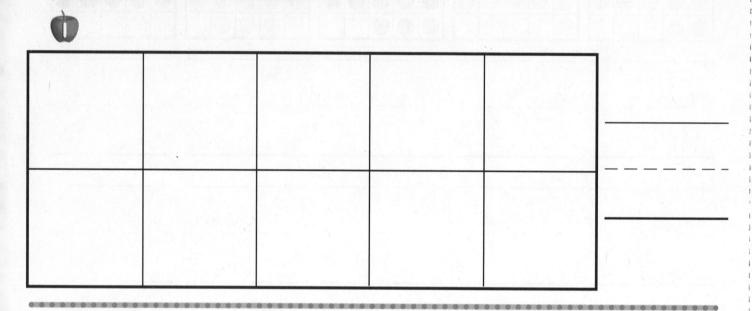

6   _____   _____

_ _ _ _ _ _ _   _ _ _ _ _ _ _

cubes

✿ **3** THINK SMARTER

7   8   _____   10

_ _ _ _ _ _ _

DIRECTIONS 1. Place counters in the ten frame to model ten. Draw the counters. Write the number. (K.CC.B.5) **2.** Use blue to color the cubes to match the number. Use red to color the other cubes. Write how many red cubes. Write how many cubes in all. (K.OA.A.4) **3.** Count forward. Write the number to complete the counting order. (K.CC.A.2)

© Houghton Mifflin Harcourt Publishing Company

Count and Order to 10

Common Core **COMMON CORE STANDARD—K.CC.A.2**
Know number names and the count sequence.

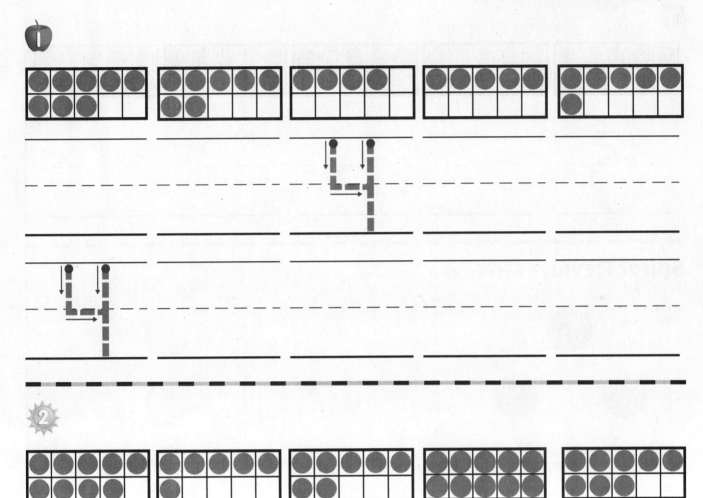

DIRECTIONS   1–2. Count the dots in the ten frames. Trace or write the numbers. Look at the next line. Write the numbers in order as you count forward from the dashed number.

© Houghton Mifflin Harcourt Publishing Company

Lesson Check (K.CC.A.2)

1

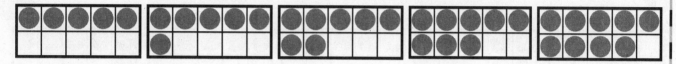

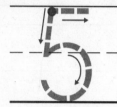

Spiral Review (K.CC.C.6, K.CC.A.3)

2

_____

- - - - -

_____

3

_____

- - - - -

_____

DIRECTIONS **1.** Count the dots in the ten frames. Trace the number. Write the numbers in order as you count forward from the dashed number. **2.** Count and tell how many are in each set. Write the numbers. Compare the numbers. Circle the number that is less. **3.** How many counters are there? Write the number.

204 two hundred four

© Houghton Mifflin Harcourt Publishing Company

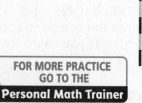

FOR MORE PRACTICE
GO TO THE
Personal Math Trainer

Name _____

Problem Solving • Compare by Matching Sets to 10

Essential Question How can you solve problems using the strategy *make a model*?

Counting and Cardinality—K.CC.C.6
Also K.CC.C.7
MATHEMATICAL PRACTICES
MP4, MP5, MP8

 Unlock the Problem

© Houghton Mifflin Harcourt Publishing Company • Image Credits: (bg) ©PhotoDisc/Getty Images

DIRECTIONS Break a ten-cube train into two parts. How can you use matching to compare the parts? Tell a friend about the cube trains. Draw the cube trains.

Chapter 4 • Lesson 5

Try Another Problem

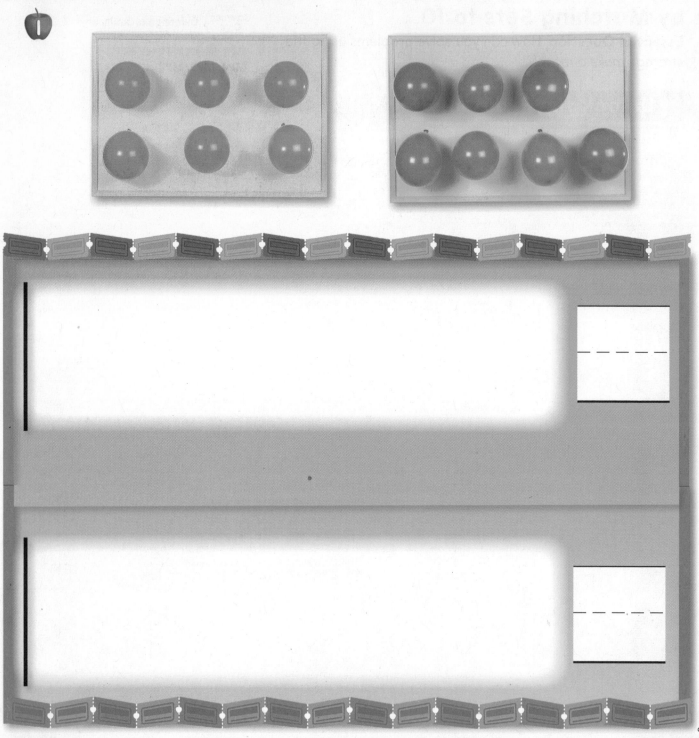

DIRECTIONS 1. Malia has the red balloons. Andrew has the blue balloons. Who has more balloons? Use *i*Tools or red and blue cube trains to model the sets of balloons. Compare the cube trains by matching. Draw and color the cube trains. Write how many in each set. Which number is greater? Circle that number.

© Houghton Mifflin Harcourt Publishing Company

Name _____

⊛2 ✓

_____

- - - - - - -

_____

- - - - - - -

_____

⊛3

_____

- - - - - - -

_____

_____

- - - - - - -

_____

DIRECTIONS **2.** Kyle has 9 tickets. Jared has 7 tickets. Who has fewer tickets? Use cube trains to model the sets of tickets. Compare the cube trains by matching. Draw and color the cube trains. Write how many. Circle the number that is less. **3.** Phil won 8 prizes. Naomi won 5 prizes. Who won fewer prizes? Use cube trains to model the sets of prizes. Compare the cube trains by matching. Draw and color the cube trains. Write how many. Circle the number that is less.

© Houghton Mifflin Harcourt Publishing Company

On Your Own

4

DIRECTIONS 4. Ryan has a cube train with red and blue cubes. Does his cube train have more blue cubes or more red cubes? Make cube trains of each color from the cubes in Ryan's cube train. Compare the cube trains by matching. Draw and color the cube trains. Write how many cubes are in each train. Circle the greater number.

HOME ACTIVITY • Ask your child to show two sets of up to 10 objects each. Then have him or her compare the sets by matching and tell which set has more objects.

© Houghton Mifflin Harcourt Publishing Company

Name _____

Problem Solving • Compare by Matching Sets to 10

Common Core
COMMON CORE STANDARD—K.CC.C.6
Compare numbers.

1

_____

– – – – – – –

_____

_____

_____

- -

2

_____

– – – – – – –

_____

_____

– – – – – – –

_____

DIRECTIONS **1.** Kim has 7 red balloons. Jake has 3 blue balloons. Who has fewer balloons? Use cube trains to model the sets of balloons. Compare the cube trains. Write how many. Circle the number that is less. **2.** Meg has 8 red beads. Beni has 5 blue beads. Who has more beads? Use cube trains to model the sets of beads. Compare the cube trains by matching. Draw and color the cube trains. Write how many. Circle the number that is greater.

Chapter 4

© Houghton Mifflin Harcourt Publishing Company

Lesson Check (K.CC.C.6)

_ _ _ _ _ _ _ _

_____

_ _ _ _ _ _ _ _

_____

Spiral Review (K.CC.C.6, K.CC.B.4b)

_____

_ _ _ _ _ _ _ _

_____

_ _ _ _ _ _ _ _

_____

_ _ _ _ _ _ _ _

_____

DIRECTIONS **1.** Mia has 6 red marbles. Zack has 2 blue marbles. Who has more marbles? Use cube trains to model the sets of marbles. Compare the cube trains by matching. Draw and color the cube trains. Write how many. Circle the number that is greater. **2.** Count and tell how many are in each set. Write the numbers. Compare the numbers. Circle the number that is greater. **3.** Count and tell how many. Write the number.

© Houghton Mifflin Harcourt Publishing Company

FOR MORE PRACTICE
GO TO THE
Personal Math Trainer

Name _____

Compare by Counting Sets to 10

Essential Question How can you use counting strategies to compare sets of objects?

Common Core
Counting and Cardinality—K.CC.C.6
Also K.CC.B.5, K.CC.C.7
MATHEMATICAL PRACTICES
MP6, MP8

Listen and Draw Real World

DIRECTIONS Look at the sets of objects. Count how many in each set. Trace the numbers that show how many. Compare the numbers.

© Houghton Mifflin Harcourt Publishing Company • Image Credits: (tc) ©Stockbyte/Getty Images (bc) ©Corbis (bg) ©Image Source Pink/Alamy

1

_____

- - - - - - - - - -

_____

- - - - - - - - - -

_____

_____

 2 ✓

_____

- - - - - - - - - -

_____

- - - - - - - - - -

_____

_____

 3 ✓

_____

- - - - - - - - - -

_____

- - - - - - - - - -

_____

_____

DIRECTIONS 1–3. Count how many in each set. Write the number of objects in each set. Compare the numbers. Circle the greater number.

© Houghton Mifflin Harcourt Publishing Company • Image Credits: (tl) ©PhotoDisc/Getty Images (tr) ©Andrew Paterson/Alamy (cr) ©Corbis (bl) ©Comstock/Getty Images (br) ©Artville/Getty Images

Name _____

© Houghton Mifflin Harcourt Publishing Company • Image Credits: (tl) ©Stockbyte/Getty Images (tr) ©Eyewire/Getty Images (bl) ©Stockbyte/Getty Images

 4

_____

- - - - - - - - - - - -

_____

_____

- - - - - - - - - - - -

_____

 5

_____

- - - - - - - - - - - -

_____

_____

- - - - - - - - - - - -

_____

 6

_____

- - - - - - - - - - - -

_____

_____

- - - - - - - - - - - -

_____

DIRECTIONS 4–6. Count how many in each set. Write the number of objects in each set. Compare the numbers. Circle the number that is less.

Problem Solving • Applications

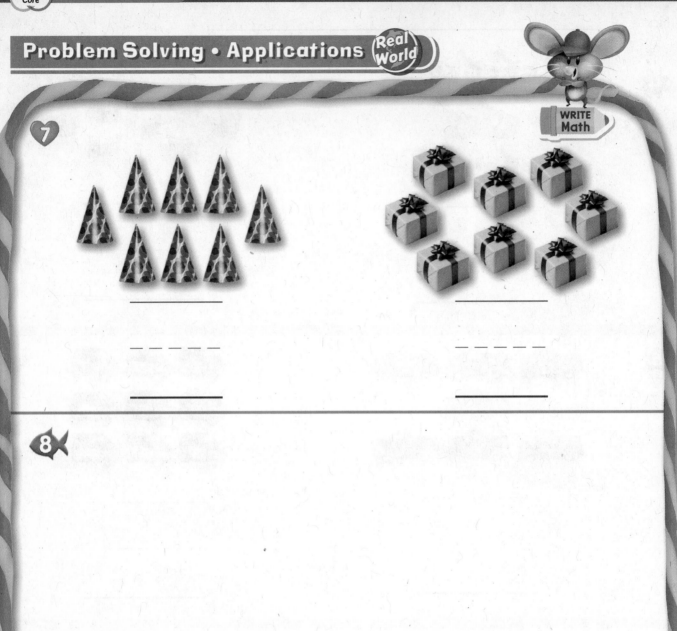

7

_____

_ _ _ _ _ _ _

_____   _____

8

DIRECTIONS 7. Megan bought hats and gifts for a party. How many hats did she buy? How many gifts did she buy? Write the number of objects in each set. Compare the numbers. Tell a friend about the sets. **8.** Draw to show what you know about counting sets to 10 with the same number of objects.

HOME ACTIVITY • Show your child two sets of up to 10 objects. Have him or her count the objects in each set. Then have him or her compare the numbers of objects in each set, and tell what he or she knows about those numbers.

© Houghton Mifflin Harcourt Publishing Company • Image Credits: (tl) ©Comstock/Getty Images (tr) ©PhotoDisc/Getty Images

Name _____

Compare by Counting Sets to 10

COMMON CORE STANDARD—K.CC.C.6
Compare numbers.

 1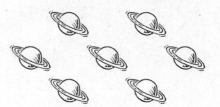

_____

- - - - - - -

_____

_____

- - - - - - -

_____

2

_____

- - - - - - -

_____

_____

- - - - - - -

_____

 3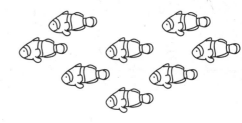

_____

- - - - - - -

_____

_____

- - - - - - -

_____

DIRECTIONS Count how many in each set. Write the number of objects in each set. Compare the numbers. **1–2.** Circle the number that is less. **3.** Circle the number that is greater.

Chapter 4

© Houghton Mifflin Harcourt Publishing Company

Lesson Check (K.CC.C.6)

1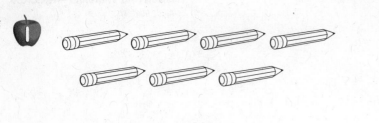

_____

_ _ _ _ _ _ _ _ _

_____

_____

_ _ _ _ _ _ _ _ _

_____

Spiral Review (K.CC.A.3, K.CC.B.5)

2

_____

_ _ _ _ _ _ _ _ _

_____

3

◯				

DIRECTIONS **1.** Count and tell how many are in each set. Write the numbers. Compare the numbers. Circle the number that is less. **2.** How many whistles are there? Write the number. **3.** How many more counters would you place in the ten frame to show a way to make 6? Draw the counters.

© Houghton Mifflin Harcourt Publishing Company

FOR MORE PRACTICE
GO TO THE
Personal Math Trainer

Name _____

Compare Two Numbers

Essential Question How can you compare two numbers between 1 and 10?

Common Core **Counting and Cardinality— K.CC.C.7**

MATHEMATICAL PRACTICES
MP6, MP8

Listen and Draw Real World

7

7 is less than 8

7 is greater than 8

8

8 is less than 7

8 is greater than 7

DIRECTIONS Look at the numbers. As you count forward does 7 come before or after 8? Is it greater or less than 8? Circle the words that describe the numbers when comparing them.

© Houghton Mifflin Harcourt Publishing Company • Image Credits: (bg) ©PhotoDisc/Getty Images

1   3   (8)

···

2   10   5

···

3   6   4

···

4 ✓   7   9

···

5 ✓   10   8

···

DIRECTIONS 1. Look at the numbers. Think about the counting order as you compare the numbers. Trace the circle around the greater number. **2–5.** Look at the numbers. Think about the counting order as you compare the numbers. Circle the greater number.

© Houghton Mifflin Harcourt Publishing Company

Name _____

6 2   4

7 5   3

8 8   9

9 10   7

10 6   8

DIRECTIONS 6–10. Look at the numbers. Think about the counting order as you compare the numbers. Circle the number that is less.

© Houghton Mifflin Harcourt Publishing Company

Problem Solving • Applications

II.

_____

- - - - - - -

_____

_____

- - - - - - -

_____

12.

_____

- - - - - - -

_____

_____

- - - - - - -

_____

DIRECTIONS **II.** John has a number of apples that is greater than 5 and less than 7. Cody has a number of apples that is two less than 8. Write how many apples each boy has. Compare the numbers. Tell a friend about the numbers. **12.** Write two numbers between I and 10. Tell a friend about the two numbers.

HOME ACTIVITY • Write the numbers I to 10 on individual pieces of paper. Select two numbers and ask your child to compare the numbers and tell which number is greater and which number is less.

© Houghton Mifflin Harcourt Publishing Company

Compare Two Numbers

Common Core

COMMON CORE STANDARD—K.CC.C.7
Compare numbers.

1

8   5

2

10   7

3

6   9

4

4   6

5

8   7

6

5   3

DIRECTIONS 1–3. Look at the numbers. Think about the counting order as you compare the numbers. Circle the greater number. 4–6. Look at the numbers. Think about the counting order as you compare the numbers. Circle the number that is less.

© Houghton Mifflin Harcourt Publishing Company

Chapter 4

two hundred twenty-one **221**

Lesson Check (K.CC.C.7)

7   8

Spiral Review (K.CC.B.5, K.CC.A.3)

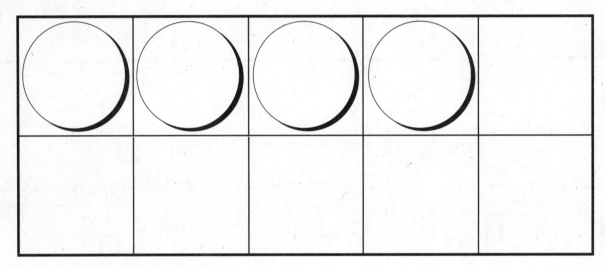

_____

- - - - - - - - -

_____

DIRECTIONS **1.** Look at the numbers. Think about the counting order as you compare the numbers. Circle the greater number.   **2.** How many more counters would you place in the ten frame to show a way to make 8? Draw the counters.   **3.** How many birds are there? Write the number.

© Houghton Mifflin Harcourt Publishing Company

FOR MORE PRACTICE
GO TO THE
Personal Math Trainer

Name _____

10

nine
ten

DIRECTIONS 1. Circle all the sets that have 10 stars.
2. How many eggs are shown? Write the number.
3. What is another way to write 10? Circle the word.

© Houghton Mifflin Harcourt Publishing Company

4

_____
cubes

_____   _____

5

3

7

Personal Math Trainer

6 THINK SMARTER +

5 6 7 8	○ Yes	○ No
8 10 9 7	○ Yes	○ No
7 8 9 10	○ Yes	○ No

DIRECTIONS 4. Write how many red cubes. Write how many blue cubes. Write how many cubes in all. 5. Look at the numbers. Think about the counting order as you compare the numbers. Circle the number that is less. 6. Are the numbers in counting order? Choose Yes or No.

© Houghton Mifflin Harcourt Publishing Company

7

_____

- - - - - - - - - - -

_____

_____

- - - - - - - - - - -

_____

8

_____   _____

- - - - -   - - - - -

_____   _____

9

7   8   9

○   ○   ○

DIRECTIONS 7. Write how many counters are in the set. Use matching lines to draw a set of counters less than the number of counters shown. Circle the number that is less.   **8.** Count how many in each set. Write the numbers. Circle the greater number.   **9.** Think about counting order. Choose the number that is less than 8.

© Houghton Mifflin Harcourt Publishing Company • Image Credits: (cl) ©Comstock/Getty Images

10
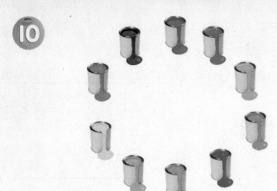

_____

- - - - - - - - - - -

_____

11 THINK SMARTER +

_____

- - - - - - - - - - -

_____

_____

- - - - - - - - - - -

_____

12

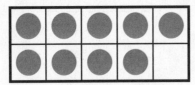   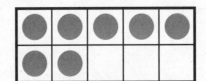   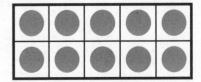

•   •   •

•   •   •

10   9   7

DIRECTIONS **10.** How many cans of paint are there? Write the number. **11.** Seth has 10 buttons. Draw Seth's buttons. The number of buttons Tina has is one less than Seth's. Draw Tina's buttons. How many buttons does Tina have? Write how many in each set. Circle the number that is less. **12.** Match sets to the numbers that show how many counters.

© Houghton Mifflin Harcourt Publishing Company • Image Credits: ©Andrew Paterson/Alamy